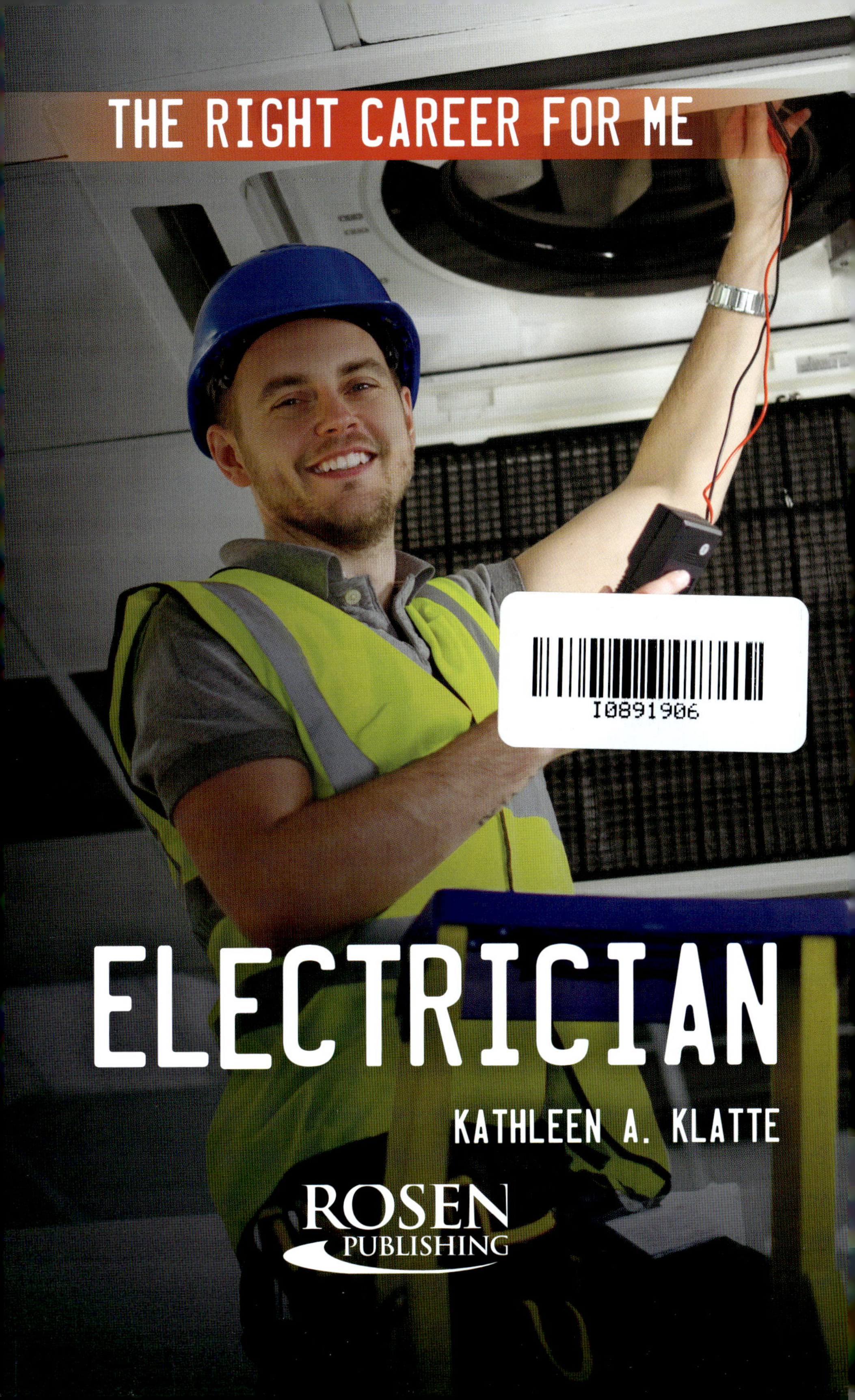

THE RIGHT CARER FOR ME

ELECTRICIAN

KATHLEEN A. KLATTE

ROSEN PUBLISHING

Published in 2024 by The Rosen Publishing Group, Inc.
2544 Clinton Street, Buffalo, NY 14224

Portions of this work were originally authored by Daniel Harmon and published as *A Career as an Electrician*. All new material in this edition was authored by Kathleen A. Klatte.

Cover Designer: Michael Flynn
Intertior Designer: Rachel Rising
Editor: Kathleen A. Klatte

Library of Congress Cataloging-in-Publication Data

Names: Klatte, Kathleen A., author.
Title: Electrician / Kathleen A. Klatte.
Description: Buffalo : Rosen Publishing, [2024] | Series: The right career for me | Includes index.
Identifiers: LCCN 2023035841 (print) | LCCN 2023035842 (ebook) | ISBN 9781499475791 (library binding) | ISBN 9781499475784 (paperback) | ISBN 9781499475807 (ebook)
Subjects: LCSH: Electrical engineering--Vocational guidance--Juvenile literature.
Classification: LCC TK159 .K53 2024 (print) | LCC TK159 (ebook) | DDC 621.3023--dc23/eng/20230811
LC record available at https://lccn.loc.gov/2023035841
LC ebook record available at https://lccn.loc.gov/2023035842

Some of the images in this book illustrate individuals who are models. The depictions do not imply actual situations or events.

Manufactured in the United States of America

CPSIA Compliance Information: Batch #CWRYA24. For further information, contact Rosen Publishing at 1-800-237-9932.

CONTENTS

INTRODUCTION

The world of the 21st century depends on electricity. We use it to preserve fresh and frozen food, wash clothing and dishes, and, of course, communicate and enjoy entertainment. Think about how many electrical systems are built right into your house. There are the lights; the heating, ventilation, and air conditioning (HVAC) system; and perhaps a security system or smart home program. Even if your home is heated by gas or oil, chances are the thermostat and ignition on the furnace are electric.

Think about what happens in a power outage. Buildings can quickly become too hot or too cold, which is dangerous for the very young, very old, or people with health problems. Food can start to thaw or spoil. Computer work that wasn't saved recently enough could be lost. People could get hurt tripping and falling in the dark. Anything that runs off an internet router—such as security systems, Wi-Fi, and some phone systems—will be offline. Home medical equipment might be running off batteries with a limited life span.

You'll need the services of a professional electrician to get things running again. If the issue is just your home, you'll need to contact an electrician to come check out your fuse boxes and wiring to figure out what's wrong and fix it.

This is the local electrician. You call her to replace a burned-out fuse or install a new light fixture.

But what if it's not just your home? What if the whole block—or the whole city—is without power? What if there are buildings with no elevator service or automated fire alarms are offline? What if there are no working traffic signals? Who fixes that? Again, the answer is an electrician—or more likely, teams of electricians.

This is also an electrician. He's checking a power substation to make sure the electricity stays on for the entire community.

You can find electricians throughout your community. They might have their own business that you call when you need something installed or fixed in your home. Electricians also work for municipalities and utility companies. Even if you don't see them, there are always electricians working hard to keep your home—and your city—safe, comfortable, and functional.

PIONEERS OF ELECTRICITY

The first documentation of electricity goes all the way back to 500 BCE and the discovery of static electricity. Over the next two millennia, there would be experiments and observations, but little practical application for electrical phenomena.

In the 18th century, American statesman Benjamin Franklin (1706–1790) undertook a serious study of electricity. He invented a battery to hold an electrical charge and described the difference between a conductor and an insulator. He also coined many of the terms we still associate with electricity—such as electrify, charge, and discharge.

The first widespread practical use of electricity came in the mid-19th century with Samuel Morse's (1791–1872) telegraph system. In less than 20 years, telegraph wires connected the United States. People could send messages by Morse code—an alphabet composed of short and long taps. A telegraph operator would then write out the letters represented by the code and deliver the message to its recipient.

The next breakthrough, generally credited to Alexander Graham Bell (1847–1922) in 1876, was the telephone. A telephone transmitted the sound of a person's voice over a wire as an electric signal. Telephones soon became the predominant method of long-distance communication in the United States.

In 1896, Guglielmo Marconi (1874–1937) invented a successful form of wireless communication—the first radio. He discovered how to send wireless messages across the Atlantic Ocean. People initially sent wireless messages in code, like telegraphs. Eventually, other scientists built on Marconi's work and figured out how to transmit music and voices.

All these discoveries eventually led to the cell phone in your pocket and the router that connects your home to the internet.

British inventors had been tinkering with electric lightbulbs since 1835. Thomas Edison made them a practical success in 1879.

Nikola Tesla (1856–1943) developed the alternating current (AC) system. This is the system that made it possible to electrify buildings.

Electricians are vital members of any construction team. They do far more than "just add lights." Today's buildings—even private homes—are built with central heating and air conditioning, as well as an array of smart appliances. Homeowners expect to install multiple computers and accessories, possibly a security system, or even solar panels. All these features might be controlled by a domotics—or smart home—system. Electricians need to be familiar with the latest technology and its system requirements.

That sounds expensive—and it is. So, what do electricians do when there's not a big market for new homes? The answer is: plenty. Electricians also work on repairs and renovations. People call an electrician to find out why that light isn't working or to install some new outlets in the garage. If they're having a kitchen or bath renovated, they might call an electrician to be sure their existing wiring can handle the new appliances. And, of course, after a couple of decades, wiring might be outdated and in need of replacement.

Becoming an electrician is a sound career choice. Although we're constantly searching for cleaner sources of power, electrified buildings and all the technology inside them aren't going anywhere anytime soon. We need plenty of well-trained electricians to keep our homes and businesses safe, comfortable, and functional.

A building project is a collaboration between architects, carpenters, plumbers, and electricians. The whole teams works together to achieve what their client wants.

THOMAS EDISON

Many of the most common electrical items you interact with on a daily basis were invented by one man. Depending on where you live, your electrical utility company might even be named after him. His name was Thomas Alva Edison (1847–1931).

As a teen, Edison was trained as a telegraph operator. He learned how the machine worked, then began inventing improvements. He later collaborated with Samuel Laws to improve the stock tickers used on the New York Stock Exchange. He earned $40,000 for that project, which he invested in a laboratory in Newark, New Jersey.

In 1876, he opened a new lab in Menlo Park, New Jersey. This is where some of his most famous inventions were born. He improved the telephone and invented the phonograph and incandescent lightbulb. The lightbulb made electric lighting practical for homes.

During World War I, Edison encouraged the government to found the Naval Research Institute. His final lab was located near his home in West Orange, New Jersey. The lab, along with his home, Glenmont, has been a national historic site since 1955.

Thomas Edison discovered the correct filament to make electric lightbulbs a practical and commercial success.

The circuit breaker that helps keep your home safe is one of Edison's 1,093 patented inventions.

CHAPTER ONE

MAKING SURE THE LIGHTS TURN ON

In basic terms, an electrician's job is to install and maintain the components that allow electricity to be used in a building. This includes things such as wires, fuses, and circuit breakers.

Electricians do many different jobs. Some install wiring in new buildings as they're being constructed. Others do repairs, renovations, and installations in existing buildings. There are electricians who work in power plants to ensure that entire communities remain electrified, and others who work in specialized facilities such as hospitals or theaters that contain a lot of electronic equipment.

The National Joint Apprenticeship and Training Committee categorizes most electricians' roles into the following four general areas:

- **Outside lineman:** This worker installs and maintains high-voltage transmission and distribution lines from power plants to homes, businesses, and factories.
- **Inside wireman:** This type of electrician installs the power and lighting systems in new office buildings and industrial facilities. The inside wireman also services electrical systems

in existing structures. They work with lower voltages than linemen.

- **Residential wireman:** Like the inside wireman, this type of electrician installs new electrical systems and maintains old ones but works primarily in homes and housing complexes. This is the electrician you call to install new lights in your yard or rewire your kitchen.
- **VDV installer technician:** This type of electrician specializes in telephone and computer systems. They must install phone circuits and computer networks according to government standards. The VDV (video/data/voice) installer technician is also an expert in security and automatic access systems, video installations, and other modern electronic designs. Telecommunications technology is a constantly evolving field. VDV technicians need to be on top of all the latest breakthroughs.

Those are merely summaries of the job descriptions of electricians. Some specialize within each category. Others perform different types of electrical work.

Linemen need nerves of steel, a willingness to work outside in a variety of weather conditions, and a strong respect for safety regulations.

Automotive electricians specialize in the electrical systems found in cars and other motor vehicles.

HAVE YOU GOT WHAT IT TAKES?

Do you think you've got what it takes to be an electrician? You need steady hands and good manual dexterity. Electricians need to be focused, detail oriented, and able to follow directions and plans precisely. Remember, everything they install will carry an electrical current. If things aren't done properly, people can get hurt or the building might be damaged.

Occupational advisers say electricians need to have a realistic personality. What does that mean? Vocational psychologist John L. Holland developed what are called the Holland Codes to help guide career seekers. He identified six personality types that can help people decide which careers might suit them best. Some of the features of a realistic personality are:

- likes to work with animals, tools, or machines; generally avoids social activities such as teaching, counseling, nursing, and informing others
- has good skills in working with tools, mechanical drawings, machines, or animals
- values practical things you can see and touch—such as plants and animals you can grow, or things you can build or make better
- sees self as practical, mechanical, and realistic

"Realistic" jobs frequently involve practical, hands-on problems and solutions—such as discovering which fuse is burnt out and replacing it. People with these jobs don't mind getting dirty.

Electricians choose tools with insulated handles to protect themselves from accidental shocks.

O*NET, a well-known provider of occupational information on the internet, describes characteristics of the electrician's work style. They need to pay attention to detail. An electrician might need to demonstrate leadership skills; take the initiative and work independently; or be a dependable, cooperative member of a team. They need to be able to adapt to a variety of workplaces and conditions. Electricians also need to be logical, analytical thinkers who can handle stress (such as time pressure during power outages and occasional criticism).

AN INTERVIEW WITH AN ELECTRICIAN

What advice can a retired electrician offer today's youth? During his 36 years as an electrician, Marion Matthews has seen many changes. Most involve computerization. "Technology has come a long way," he says.

Matthews became an electrician's apprentice after serving in the U.S. Navy. "I didn't have any marketable skills," he says. "I'd always wanted to be an electrician. It just amazed me how electrical systems work. To bring power to homes, make lights burn, provide heating—that's what I wanted to do."

His apprenticeship involved on-the-job training and twice-a-week classes. It lasted four years. Afterward, he became a journeyman electrician under union contract. Although he lived in South Carolina, project assignments took him coast to coast. He worked in houses, schools, factories, paper mills, shipping yards, and nuclear power plants. As construction methods and technologies changed, he and his coworkers constantly learned new skills and techniques.

Matthews advises young people not to consider this career path unless they have an aptitude for math and a keen concern for safety. "Electricity is magical, but it can kill you if you aren't careful."

Recently retired, Matthews looks back on his career with deep pride. "It was a good way to make a living, and it's a great field to get into. You help improve other people's lives."

GETTING THE JOB DONE SAFELY

Most of an electrician's work falls into one of two categories. They install the wiring and electrical fixtures in new structures. They also rewire, repair, and renovate electrical elements in old buildings and in remodeling projects. There are many different types of structures. They include homes, outbuildings such as workshops and detached garages, office buildings, factories, hospitals, college buildings, and military and other government installations. All of them must be wired properly.

The wiring of a new project begins with planning. An electrician or electrical team studies blueprints of the building and makes diagrams of the wiring that is needed. Electricians note where fixtures and outlets are needed and how the wiring should be connected. Electricians then run the necessary cables and wires throughout the facility, including underground cables. They install light fixtures and dimmers, receptacles, and switchboxes.

They run connecting wires through wall spaces. They provide special sockets for heavy appliances, such as refrigerators and clothes dryers. They make sure the wires and circuits can safely accommodate all the equipment intended to run in each area of the building.

The building project may call for an electrician to install yard lights or streetlights. Depending on the type of structure, there might be security cameras or lights that are activated by motion.

When a power outage occurs, an electrician must find the cause and solve the problem. A facility like a hospital where lives depend on electrical equipment functioning properly might have electricians on staff who routinely inspect and test electrical installations to try and catch problems before they occur.

Some electricians are on staff in other types of large buildings. They make sure each system is running safely and requires as little energy and expense as necessary. They troubleshoot and

Electricians often work on remodeling projects. Sometimes people want to move major appliances to a different part of the room. Other times, a worker discovers old or damaged wiring that needs replacement.

test fuses, circuit breakers, transformers, and the other parts of electrical systems. No one wants to have people stuck in an elevator or see a business lose a day's worth of data because some routine system failed.

A homeowner generally becomes aware of a problem when they flip a switch—and nothing happens. Then they'll call a local electrician to come out and figure out where the problem is and fix it.

SAFETY CODES

Electricity is an amazing force. It lights and heats our homes, keeps our food fresh, and makes most of today's business possible. But electricity can be dangerous if it's not handled properly. Overloaded circuits or damaged wires can cause fires. Power surges can damage computers and other electronic equipment, and people can die from electrical shocks.

Safety codes exist to keep people and property safe. Every municipality has building codes. Any new construction and most major renovations must be thoroughly examined by a building inspector before people can occupy the structure.

Electrical work needs to comply with the National Fire Protection Association's NFPA 70, National Electric Code (NEC). All 50 states have adopted this code. It describes the standards for design, installation, and inspection of electrical work.

The Occupational Safety and Health Administration (OSHA) also has standards designed to keep people safe when working with electricity. This includes standards for personal protective equipment (PPE) that must be worn when doing electrical work.

A steadfast rule for all electricians is to make sure all work complies with codes and regulations. City, county, state, and national government departments issue codes. Certain industries impose additional standards for electrical work. Some facilities or types of equipment might require battery backups to ensure uninterrupted operation. Regulations vary from state to state around the country. The electrician needs to thoroughly understand the regulations that apply to every project he or she takes on.

Work on-site is only part of a professional electrician's job. Accurate paperwork is required in any business. A solo electrician or the owner of a small company has a business to run. An electrician working in the field must report on every job completed. The electrician's paperwork includes recordkeeping, writing reports, filling out insurance forms, placing equipment orders, and managing the firm's billing and banking.

Electricians need to keep accurate records of their work. They need to record what parts and supplies they used on a project so they can bill their client correctly and order more for the next job.

An independent electrician also needs to have good communication skills. They might need to explain to a homeowner that they're doing things to make the building safer and more functional. For example, they might need to install more electrical outlets or replace old, damaged wires. They're not doing this to add to the bill, but because it's required by the local building codes.

SOMETIMES IT'S A MATTER OF LIFE AND DEATH

Sometimes, keeping the power flowing isn't just a matter of keeping the lights on or preserving business data. Sometimes, people's lives depend on it.

Jerry Van Liew, an electrician in Charleston, South Carolina, began learning the trade when he was a teenager. During summers, he helped his grandfather. Van Liew's granddad was a residential electrician who wired new homes and home additions and made service calls to older houses. Van Liew, however, took a different career path. He went to work on the electrical staff of the Veterans Administration Hospital in Charleston. In time, he became the hospital's electric shop supervisor.

Why does a hospital need a whole staff of electricians? Electricity is vital at a large facility such as the VA hospital. It has 117 beds and serves more than 40,000 patients each year. An equipment failure during surgery or while a patient is on life support can be fatal.

"A typical workday for the electricians at our hospital involves mostly upkeep," Van Liew explains. "We have a lot of equipment. It requires daily checks and preventative maintenance."
He shares these examples:

- checking every item of electrical medical equipment in the operating rooms every day
- replacing lightbulbs
- adding electrical sockets, lighting, and special computer jacks in various offices
- checking fire alarm and nurse call systems
- maintaining security surveillance systems
- ensuring that emergency backup generators constantly are in working order
- maintaining television sets in patients' rooms and lobbies
- maintaining computers and phone systems at staff workstations

A DAY ON THE JOB

What do electricians do all day? That depends on what their specialty is. An electrician who is hired to install the wiring in a new home or commercial building usually begins by studying the builder's blueprints. Blueprints are a special type of drawing that shows where everything is supposed to go as the structure is built, including plumbing and wiring. Blueprints indicate the location of the central electrical system controls, wiring circuits, wall sockets and switches, lighting, and major appliance space.

Electricians then begin installing the system, component by component. They use a variety of tools: hand and power saws and drills, wire strippers, conduit benders, cutting devices, pliers, and screwdrivers. After connecting the elaborate scheme of wiring to circuit breakers and transformers, they must test each connection.

Testing tools include:

- voltmeters, to measure voltage
- ammeters, which measure electric current in amperes
- ohmmeters, which measure electrical resistance, measured in ohms

Electricians who work mainly in larger construction projects (industrial buildings and office complexes) have more heavy-duty electrical components to work with. They pull groups of wires and cables through large conduits that are concealed between walls. Some are trained to install special cables that will bring power to major telecommunications systems.

Typical jobs are different for electricians specializing in maintenance and repair. Their main task is to identify problems and potential trouble spots in older buildings. Older buildings might need rewiring to accommodate updated technology, such as high-speed internet which is often used for streaming content on televisions as well as for business and communications.

A notable difference between installers of new systems and repairers of old ones is the nature of the time pressure. An electrician who wires a new structure works under the schedule of the general contractor. The building goes up methodically and the electrician is called for at certain stages of overall construction.

When a repair professional is needed, on the other hand, the job often is an emergency. An outage may have left a family or business without power. The affected people may include individuals with serious health problems. Businesses may suffer huge financial losses if they can't return to normal quickly. Sometimes an outage may indicate a potential fire hazard. Customers expect the service professional to arrive as soon as possible and troubleshoot and fix the problem quickly.

Some electricians perform inspections and maintenance. They usually follow periodic schedules. They might plan their site visits well in advance and arrive with a checklist to follow. Inspectors look for potential problems in electrical equipment and circuitry.

Maintenance workers respond to reports of things that aren't working properly or need replacing. They also have checklists of fuse boxes, circuit breaker boxes, lighting, and other parts of a system that they check regularly to prevent problems from occurring. Those who work in factories and other large buildings may service motors and heavy equipment. Many large institutions keep electricians on the job around the clock, working in shifts.

Maintenance workers also function as advisers. They report their observations to the building manager or owner so that potential problems can be corrected. They might suggest replacing or updating equipment before it malfunctions. They point out equipment that is beginning to deteriorate to plant managers before there's an accident. They explain the dangers and other possible consequences. This might include failing a government inspection, which could lead to a fine or loss of insurance.

No one wants to hear that they need to make a major repair, but part of an electrician's job is to find and fix things that aren't up to code before a fire happens or someone has an accident.

CHAPTER TWO

CHOOSING A SPECIALITY

Often, a construction project requires more than one type of electrician. One team might install all the wiring in a building, and another may connect it to the outside power grid. Specialists might be called on to install a computer network, video monitoring system, alarm system, home entertainment system, or a smart home system to run everything. Some of the many types of electrical professionals are described in this chapter.

ELECTRIC METER INSTALLER AND REPAIRER

Power meters automatically record how much electricity a home or business uses. The meter reading determines what the monthly electric bill will be. Usually, the company that supplies power supplies the electric meter. Specialists who work for the electrical utility company will install meters outside homes or office/factory buildings, or on connecting power poles. They test the power meters and repair them when necessary.

In the past, utility company employees went door to door every month or quarter to visually take a reading from the meter. Today many electrical utility companies have switched to smart meters. Smart meters transmit data directly to the utility company. Customers can check their usage and preview their bill on the company's website.

This is an ohmmeter. An ohm measures electrical resistance.

ELECTRICAL ENGINEERING TECHNICIAN

Electrical engineering technicians work with engineers to develop new equipment that's powered by electricity. They work in offices, labs, and factories. They help design, build, and test new equipment to solve specific problems. They might work on medical devices or communication equipment. Sometimes they might visit a site to be sure a new system is working properly and fulfilling the task it was designed for. This is a job for people who are creative problem solvers.

THE LIFE OF A LINE WORKER

An electric company "bucket truck" stopped at the side of the road is a fairly common sight. The professionals up in the bucket are called line workers. It's their job to make sure that electricity is flowing through the heavy wires—or lines—that connect to each house and business.

The invention of power-lift buckets has made the line worker's life easier and safer. Before they became common, line workers relied on spiked shoes and safety belts (and no small amount of muscle and acrobatic skill) to climb poles. They are still trained in pole climbing because buckets are not practical in every situation. It's a skill that's practiced in line workers "Olympics"—competitions where they need to perform a skill or set of skills. They're judged on how quickly and effectively they complete each task.

Line workers must be able to perform physically demanding work at significant heights, often for long periods. Not surprisingly, an essential item of the line worker's equipment is the hard hat. They also wear gloves, high visibility vests, and rubber-soled boots. Line workers also know how to install and repair underground power cables. Depending on the task at hand, line workers may work solo or in pairs or crews.

Besides being able to "work with the wires," this professional must be able to operate the truck and each of its devices. The line worker must be skilled in excavating and trenching. Safety training is required. In case of an accident, an injured worker's first source of assistance will likely be a coworker. Line workers are trained in pole-top rescues and other peculiar types of rescue. Sometimes these rescues are part of their Olympics. They can also perform life-saving first aid, including cardiopulmonary resuscitation (CPR) for victims of electrical shocks.

ELECTRICAL EQUIPMENT TESTER AND INSPECTOR

Knowledgeable inspectors and testers make sure specific types of electrical systems are working properly. They may be trained in inspecting and testing computer systems, for instance, or broadcast station transmitters, or radar/SONAR devices. Inspectors and testers must be familiar with the NEC as well as the standards of the particular type of equipment they're examining. They also conduct inspections in factories. They must be skilled in using special measuring tools and be meticulous record keepers.

ELECTRICAL REPAIRER

Companies, industries, powerhouses, and substations rely on complicated electrical installations. Some run heavy electrical equipment such as assembly lines or electronic transmitters with controls and antennas. Many of the systems operate around the clock. Electrical repairers make sure when a breakdown occurs, it's fixed in very short order. Industrial repairers troubleshoot problems and make repairs. They work with everything from computer diagnostic programs to screwdrivers and pliers. Any repair must be carefully logged. Even when there is no problem, many companies keep repairers on staff. They install new electrical products, test old systems, and make any needed adjustments.

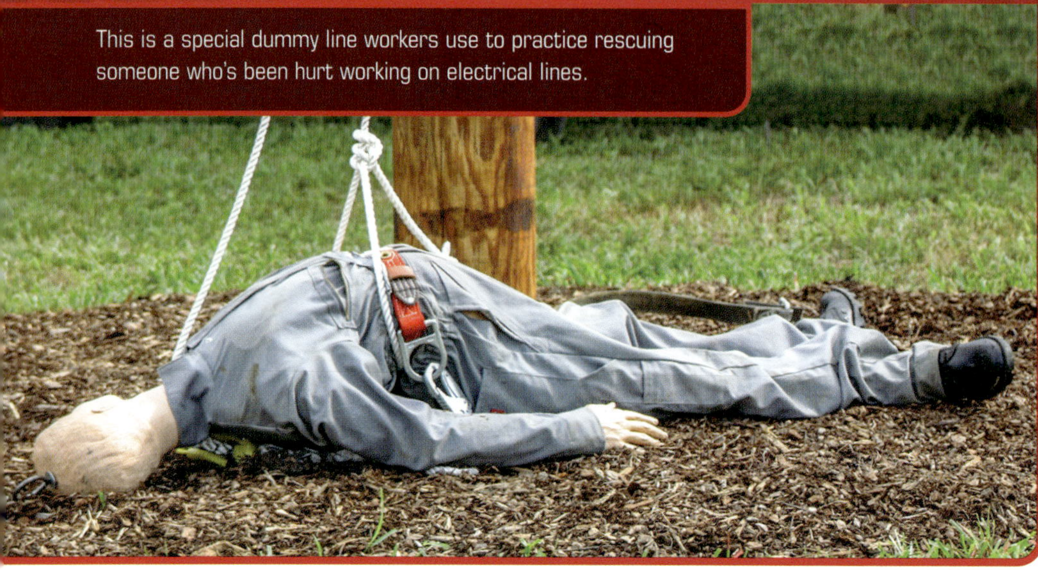

This is a special dummy line workers use to practice rescuing someone who's been hurt working on electrical lines.

POWER LINE INSTALLERS AND REPAIRERS

While some industries might have backup generators for emergencies, homes, offices, and factories do not generate the electricity they use on a regular basis themselves. They draw electrical current from an outside provider. Their electric company generates electricity via fossil fuel, nuclear power, wind turbines, solar panels, or hydroelectric dams. The companies transmit power from stations to substations and finally to individual buildings.

Line installers and repairers keep each customer connected to the outside power company. They repair and replace wires and cables for existing customers and install lines from the street to customer locations. These workers may be employed by the power company or by a separate electrical firm the power company engages to do the work. Some power line workers erect power poles and even transmission towers.

WHAT'S NEXT?

Electricians don't have to spend their entire careers installing and repairing things. Eventually, they might want to find work that doesn't involve so much heavy lifting, being out in the weather, or tracing wiring through basements or attics.

One option is to go into the electrical parts retail business. A professional electrician or home handyman who enters an electrical parts shop expects to deal with a salesperson who knows ohms from amps. A novice will appreciate help from someone who's familiar with different brands of tools and can recommend the best value. An experienced electrician behind the counter is able to "talk shop" comfortably with any customer and ensure consumer satisfaction. That brings customers back—and generates new business by word of mouth from happy buyers.

A busy professional will appreciate being able to find exactly what they want quickly. A novice or do-it-yourselfer will be glad of someone who can steer them to the proper tools and supplies.

A different option is teaching. Some professional electricians possess special people skills in addition to their "realistic" get-it-done personalities. They might consider furthering their education and becoming certified to teach at high schools, vocational institutions, colleges, or other training centers.

THE ENTERTAINMENT INDUSTRY

When you think of the theater, one of the first things that comes to mind is probably lights—brightly lit marquees on Broadway, a spotlight on the soloist for their big number, or colored lighting setting the mood for a scene. Or you might think of the microphones and speakers that help you hear every word and note with perfect clarity. Or what about those cool on-stage special effects?

All these things involve electricity, so it shouldn't surprise you to know that there are electricians who work in theaters, movie studios, and other live event spaces.

Theater electricians are responsible for installing and maintaining the electrical equipment—generally for sound and lighting—used in a production. A person called a lighting designer decides how they want the stage to be lit. The electrician then installs the correct fixtures and bulbs to produce the desired effect. Theater electricians keep everyone in the theater safe by making sure that lighting equipment is properly secured (so it won't fall), and that circuits aren't overloaded and lights are fitted with the correct wattage of lightbulbs (to prevent fire hazards). They might also be responsible for setting up and dismantling work lights used for building and striking a set.

Theater electricians are represented by the International Alliance of Theatrical and Stage Employees and Moving Picture Technicians, Artists and Allied Crafts of the United States, Its Territories and Canada (IATSE).

Imagine your favorite band trying to play onstage with no lights or microphones! Theater electricians make sure you can see and hear all the action. They also make sure that all their equipment is safe from shocks and fire hazards.

WORKING CONDITIONS

Just as job types vary, so do working conditions. Electricians in different jobs have different tasks to perform. From time to time, most must work in situations that are unpleasant. They may need to wear protective gear—hard hats, goggles, special shoes and gloves, and earmuffs.

Wires in an office may be run across a dropped ceiling. To access the wires, an electrician might have to climb a ladder and move ceiling tiles. Wires in a house might be in a basement, crawl space, or attic. Electricians frequently must work in cramped quarters. They may have to twist their bodies into awkward positions for long periods. They might encounter mold, mildew, or rodents.

Besides the expected hazard posed by electrical currents, electricians may also encounter material hazards. Asbestos, for example, was a type of insulation commonly used around heating systems until the late 1900s. Its advantage is that it is heat-resistant, so it can help reduce the risk of fire. The construction industry phased out the use of asbestos after studies linked it to lung disease. However, you can still find it in millions of older homes and office buildings. Electricians who repair and inspect wiring in older buildings should have the proper PPE to protect them from mold spores, asbestos, and fiber glass particles.

Those who work outdoors, such as power line installers and repairers, have their own worries—particularly the weather. Severe weather can cause power outages and they need to get to work restoring power as quickly as possible—sometimes while the conditions are still unsafe for most people.

Winter storms bring down trees, power lines, and power poles. Workers face icy elements to restore power to customers as quickly as possible. They need boots and gloves for traction on icy surfaces, and warm outerwear to protect them from freezing temperatures. They need to be aware of the danger of frostbite. They take these risks because they know very young children and elderly people are in life-and-death situations without heat.

In the summer, heat waves can cause power outages. Air conditioning systems require a great deal of power, and when too many are on at the same time, the electrical grid might not be able to handle the strain. Workers often toil for hours in sweltering temperatures to retore power.

An ice storm brought down a power line. Fire department and other emergency services will always wait for a crew from the electrical utility company to shut down power before they approach an accident scene.

Electrical storms pose obvious hazards. Besides electrical shock, workers sometimes are injured in falls and suffer muscle strains. Line workers have special training to aid one another in case of an accident.

In most jobs, electricians have a typical workweek. They often have to work overtime, though, and they may be called on at odd hours for emergency service. In wide-scale emergencies such as ice storms, hurricanes, or heat waves, teams of electricians may work for days and nights on end with only brief breaks.

AN ELECTRIC VOCABULARY

Electricians use many special words you may not have seen before. They also use common words in different ways.

AMPACITY	The amount of current that can safely be transferred through a wire.
AMPERE	The base unit of electric current.
CIRCUIT	The complete path that an electric current travels along.
CONDUCTOR	A material such as a metal wire through which electricity flows freely.
FEEDERS	Power lines through which energy is transmitted.

FUSE	A safety device to prevent fires caused by overloads.
INSULATOR	A material wrapped around wires to prevent electrical shocks.
KILOWATTS	The rate at which something uses energy.
RACEWAY	An enclosure that protects a group of wires.
RESISTANCE	The measure of how difficult it is for electricity to pass through a substance.
VOLT	A measurement of electrical potential, force, and difference.
WATTAGE	A measurement of electrical operating power.

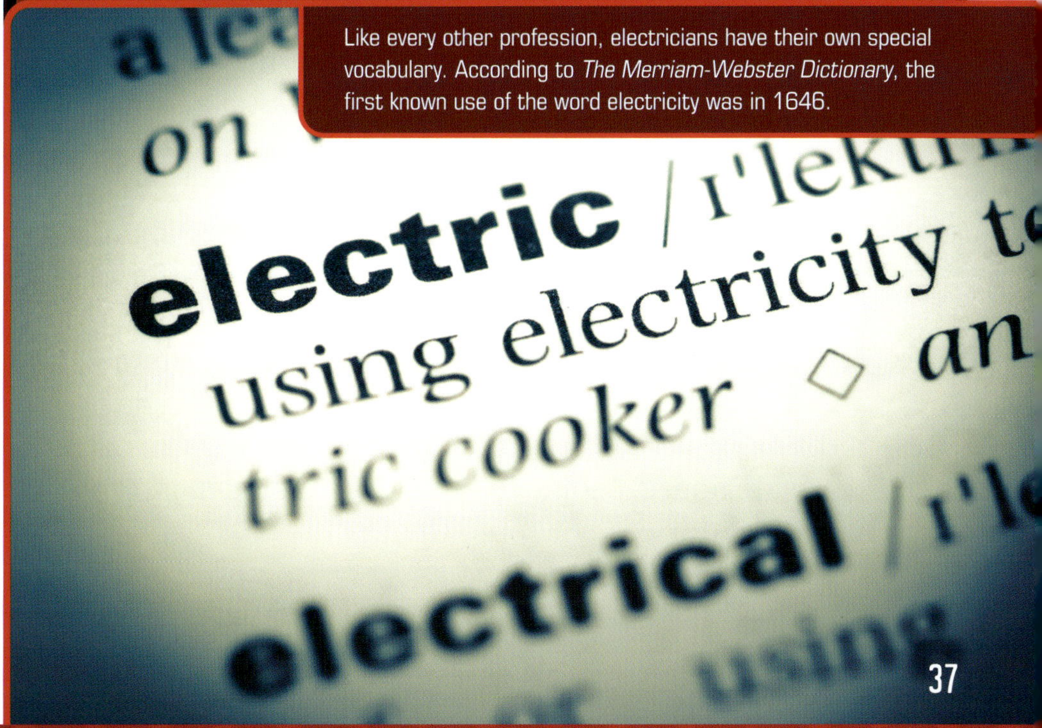

Like every other profession, electricians have their own special vocabulary. According to *The Merriam-Webster Dictionary*, the first known use of the word electricity was in 1646.

CHAPTER THREE

HOW DO I GET STARTED?

An electrician needs to be a practical problem solver. They need to understand their own job and how to do it efficiently and safely. But they also need good people skills. An electrician needs to be able to communicate effectively with other professionals on a building project. They may need to coax information or explain something to a customer who doesn't understand technical language.

PART OF THE BIG PICTURE

A professional electrician needs to understand not just how a building is wired but how it is put together and how the other systems work. They need to understand the intended use of the building they're working on. Are there enough outlets in an office space to accommodate all the computers, printers, and copiers without resorting to power strips or extension cords? Or what about a hospital room where a patient's life might be jeopardized if a complex piece of equipment loses power? Even in a private house, people today expect to plug in more electrical devices per room than in the past.

They should know the qualities of different building materials. Especially, they should automatically recognize materials that are flammable or that pose chemical hazards. They must understand home and office design. Electricians consult building blueprints and draw their own wiring circuit schematics. Their wiring must conform to the overall blueprint; architects don't design buildings around the electrical circuitry.

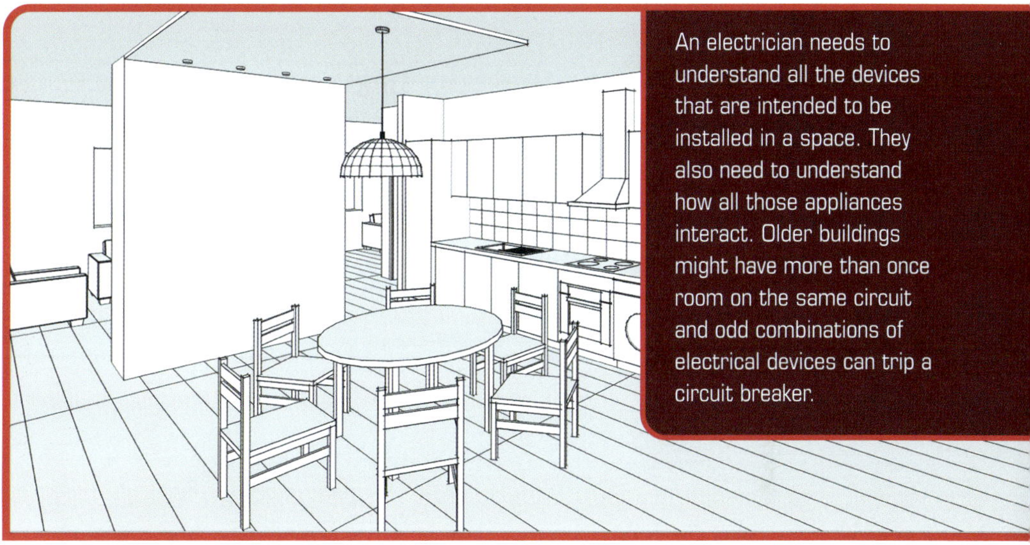

An electrician needs to understand all the devices that are intended to be installed in a space. They also need to understand how all those appliances interact. Older buildings might have more than once room on the same circuit and odd combinations of electrical devices can trip a circuit breaker.

SMART AND SAFE

An electrician must thoroughly understand every aspect of installations and repairs. When they begin their training, they'll learn about the types of electrical current. Power measurements (volts, amps, watts, ohms) will become as familiar as ABC. Electricians know why different types and sizes of wires and cables are needed for specific purposes. They know how much power various types of electric motors need.

Most important, they understand safety practices. The first safety rule is to disconnect the current before working on an electrical circuit or component. Even the fire department will defer to an electrician before touching any kind of power line. Touching a "live wire" can shock and even electrocute the worker. Proper grounding is also essential to prevent electrical disasters. Electricians also learn to spot fire hazards instantly.

A career as an electrician doesn't require a college degree—many attend a technical school instead. However, they do need one academic strength: a good understanding of math. In order to advance in their careers, electricians must understand not only basic arithmetic but also geometry, algebra, calculus, and statistics.

Even if the power has been turned off, an electrician will still check with a voltage tester before they touch anything. Better safe than sorry!

SOLVING ELECTRICAL MYSTERIES

While electricians should have what career counselors call "realistic" personalities, they also need good personal skills, especially when it comes to listening to clients. Sometimes they need to combine their problem-solving skills with "thinking outside the box."

Many electrical tasks and problems are simple to deal with: installing a ceiling fan, replacing a burnt fuse, etc. Some safety hazards are obvious—like that gizmo from the discount store that allows you to plug six things into one outlet.

Others are much more difficult. Why does an electrical circuit in a home seem to regularly overload? Is there more than one room on the circuit? How old is the wiring? Or is there a defective socket? The electrician gathers all the information about the problem from every source available. The worker considers the possibilities and decides on the first, best approach to solving the problem. This often calls for good judgment as well as experience and knowledge.

LOGIC VERSUS PEOPLE

Electricians are precise, logical thinkers. They understand the capacity of wires and circuits and install the appropriate equipment for a structure. But once people are living and working in a space, things can become a bit illogical. An electrician might survey a structure and immediately understand what's causing a problem, but they need to be able to express that in terms the customer can understand.

For example, most people understand that large appliances such as refrigerators or washing machines need to be professionally installed, need special heavy-duty sockets, and might even need a dedicated electrical circuit. But they might not understand how much current that little beer fridge or space heater draws, or the difference between a surge protector and a cheap multi-plug accessory.

An electrician needs to be able to clearly communicate why something is nonfunctioning or unsafe and suggest solutions.

An experienced electrician becomes a detective of sorts. They develop a "sixth sense" in their work. As they explore a job site, they learn to detect problems and potential problems that are not obvious at first glance. O*NET, the online provider of occupational information, calls this trait "problem sensitivity."

O*NET points out that an electrician needs good "deductive reasoning" and "inductive reasoning" powers. Deductive reasoning is "the ability to apply general rules to specific problems to produce answers that make sense." Inductive reasoning combines bits of information "to form general rules or conclusions"; this "includes finding a relationship among seemingly unrelated events."

For example, why does a home always have electrical issues when they have house guests? Is it just Murphy's Law? Or is there a logical reason? Perhaps it's the two hair dryers being turned on at the same time in the guest bathroom. Or maybe it's the other current-drawing devices being used simultaneously in guest bedrooms—all on the same circuit. The electrician can explain to the homeowner how to remedy the problem by taking turns using appliances. Or maybe it's time to upgrade the wiring.

As the electrician gains experience, knowledge accumulates. The electrician learns to weigh the strengths and risks of different approaches that could be taken in any given situation. Time turns apprentice and journeyman electricians into master electricians—"experts."

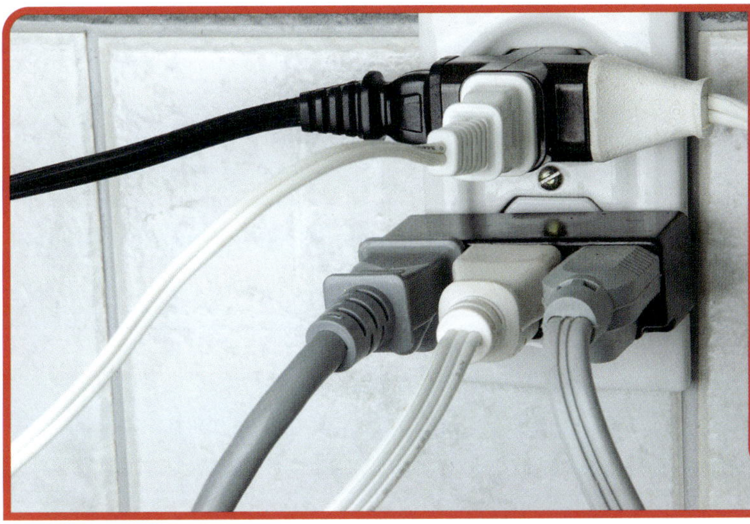

It's easy to find cheap gadgets that allow you to "add more plugs." Generally, these are unsafe. It's much better to speak to an electrician about safely increasing the electrical capacity of a space.

A surge protector has a three-prong grounded plug. It's designed to protect your electrical equipment from power surges. You can buy them in office supply or home improvement stores. A salesperson can recommend the best model for the devices you plan to plug in.

UL SOLUTIONS

An electrician studies and serves an apprenticeship to learn how to work safely with electricity. But how does a member of the general public know that the thing they're plugging into the wall is going to function properly and not start a fire? One way is to look for a seal with a capital U and L in a circle. This is the mark of UL Solutions, formerly known as Underwriters Laboratory.

UL Solutions is an independent testing lab. That means they don't work for any manufacturer. They conduct rigorous testing on products to ensure they're safe and work the way they're supposed to.

THE RIGHT STUFF

Electricians need to be in good shape. Their work is often physically demanding. Working on a new building can mean carrying heavy reels of cable or ladders. Renovations and repairs might entail opening walls or squeezing into small spaces.

Keen eyesight is vital. One reason is that electrical wires are color-coded for safety. Some colors are dictated by NEC regulations. Others are industry standards. Another is that sharp near vision can mean the difference between a successful job and calamity. Electricians frequently find themselves studying wire connections from a distance of only a few inches.

Good hand-eye coordination and manual dexterity are essential as well. An electrician must be able to move their hands and arms quickly and work nimbly with their fingers and thumbs. They need to grasp and manipulate tools, wires, and other objects. Situations can quickly become uncomfortable. After a few minutes at an awkward body angle under a crawl space, they may want to adjust their body position while keeping a grip on the pliers or screwdriver.

Finger dexterity likewise is crucial. Many types of electrical screws are tiny. Capping or splicing procedures often involve delicate wire ends.

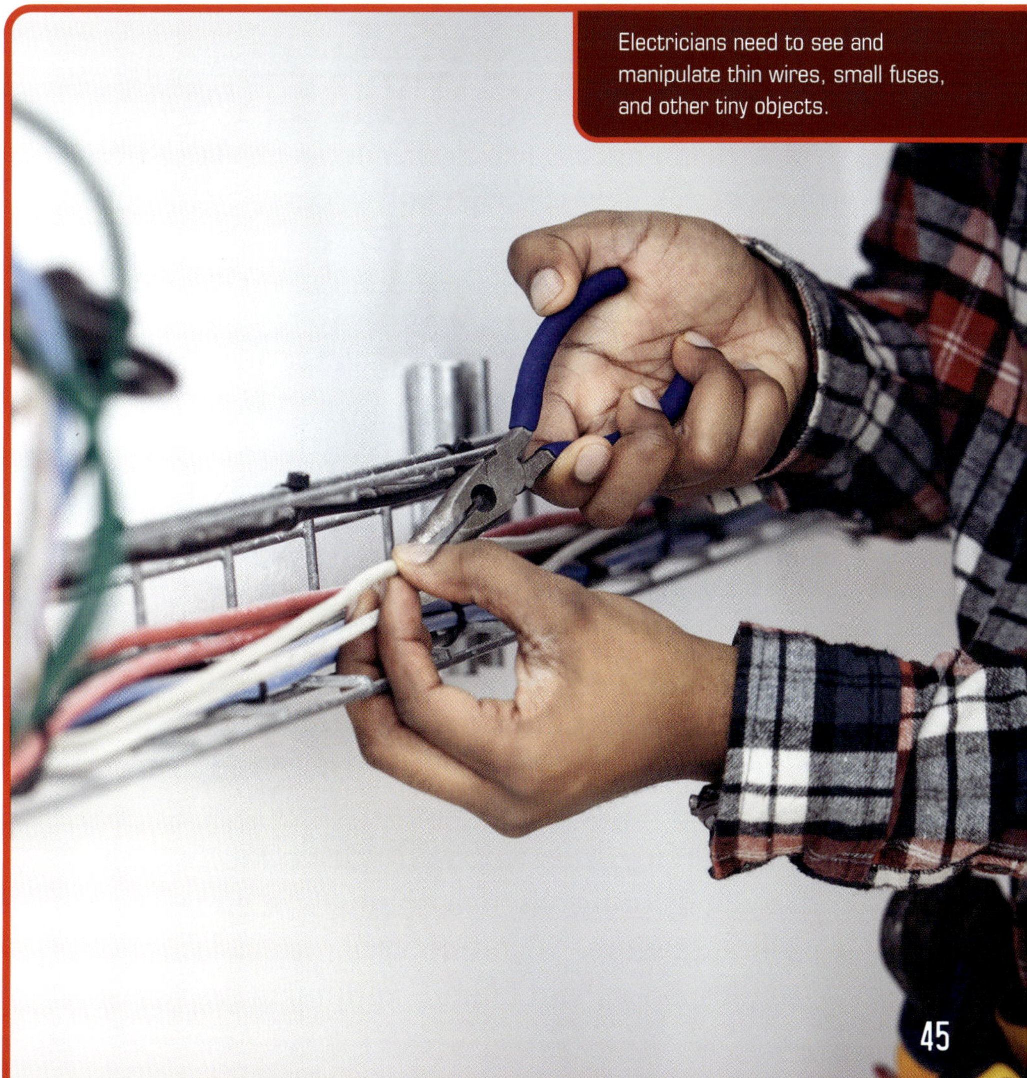

Electricians need to see and manipulate thin wires, small fuses, and other tiny objects.

Indoors or out, the work sometimes is strenuous. Electricians may have to lift and carry heavy equipment. They may work at heights, standing on ladders or scaffolds. They must crawl on hands and knees and crouch in uncomfortable positions—sometimes for extended periods of time—as they install, replace, and repair wiring, fixtures, and other items. They may need to climb poles or go up in the "bucket" to fix something. They need an excellent sense of balance and no fear of heights.

WHAT'S IN YOUR TOOLKIT?

An electrician must be good with tools. Every utility professional knows the secret of success is in having the right tools for the job—and knowing how to use them. Through experience, an electrician will discover favorite brands of tools and supplies and learn the best places to purchase them. Besides screwdrivers and wire cutters, what items might be found in the electrician's toolkit?

- **Stripping devices:** These include sharp grips that can strip synthetic insulation off bare wires (in order to make new connections) and other implements. These can be found at big box home improvement stores in a range of sizes and prices. For electrical work, they need to have insulated handles.
- **Measuring and testing devices:** Electricians frequently bring out their ohmmeters, ammeters, milliammeters, voltmeters, test lamps, oscilloscopes, and other equipment to check

current, intensity, etc. Their main concern is to be sure the electrical system or machine is functioning safely. These can also be found in big box stores, but a novice might be better off in a specialty store where they can get expert advice.

- **Power tools:** It's often necessary to use saws, drills, and other power tools to perform an electrical installation or get to a behind-the-wall electrical problem.
- **PPE:** Outdoor work requires gloves, high-visibility vests, and hardhats. But a professional should be prepared for indoor hazards as well. Safety glasses, a variety of dust masks, and earplugs or noise canceling earmuffs are always good to have on hand.

This electrician is wearing proper PPE for working on a construction site.

- **Wire and cable reels:** Enormous reels contain large quantities of cable needed for construction projects. Heavy duty stands and jacks keep the cables and wires orderly and make it easy to dispense what's needed.
- **Computer software:** Electricians rely on computers as much as office workers do. They use many kinds of software to do their work. Analytical software helps then evaluate electrical and lighting installations. Architectural and design programs help them plan and diagram a new electrical installation throughout a building. Project management and database software helps them make precise job estimates and keep accurate records. Electricians who run their own shop might use standard office programs to communicate with clients and track expenses.

Some electricians may be involved in setting up computer installations. These might include security systems, domotics or smart home suites, or even systems that control the functions of large buildings.

Aside from knowing how to use their tools and equipment to do their jobs, electricians should know how to store it properly. They should also be able to perform spot repairs. It's not always possible to stop work and run to a store for a replacement. Being able to make even a temporary repair can help keep a job on schedule.

According to the Bureau of Labor Statistics, 2 percent of American electricians are women. They make up 8.2 percent of the membership of the International Brotherhood of Electrical Workers.

CHAPTER FOUR

EDUCATION AND TRAINING

To begin studying to be an electrician, candidates generally must be 18 years old and have a high school diploma or General Educational Development Test (GED). In some states, 16- to 17-year-olds might begin training with parental approval.

Students who'd like to pursue a career as an electrician should have a solid background in math and science. Physical education (PE) and a sport are also helpful, as the job can be very physical. Extracurricular activities involving electronics, such a robotics club, will look good on a transcript. See if your school district offers any sort of industrial art or career technology programs. Anything to do with carpentry or construction can also be useful, as electricians work on construction projects.

Training programs for electricians involve both classroom instruction and work supervised by professional electricians. For example, an inside wireman apprenticeship takes five years. It requires about 2,000 hours of paid on-the-job training (OJT) per year as well as several classroom hours per week. Some programs require a certain high school grade point average (GPA). They may also require specific math classes, either in high school or from college.

Single classes can often be taken through continuing education programs at local colleges.

Because apprentices are employees, they're subject to the requirements of the business where they're apprenticed. This might include taking a drug test, proving that they can lift a certain amount, or having a valid driver's license.

In most scenarios, the trainee then must obtain a license by passing an exam. The applicant must demonstrate a working knowledge of electrical theory and an understanding of national and local codes—particularly, the National Electrical Code (NEC). An Occupational Safety and Health Administration (OSHA) test may also be required.

Apprenticeships are administered by training committees. These committees might consist of members of the International Brotherhood of Electrical Workers (IBEW) union, the National Electrical Contractors Association (NECA), other electrical associations, and local electrical company professionals.

The National Joint Apprenticeship and Training Committee (NJATC) offers apprentice training, journey worker training, and instructor courses at training centers throughout the United States and Canada. NECA and IBEW develop the courses. Apprentice training is available for those who want to become outside linemen, inside wiremen (commercial and industrial buildings), residential wiremen, and VDV installer technicians. Journey worker courses teach new skills to veteran electricians as new technologies are developed. Advanced training is available for qualified electrical instructors.

The first step to an apprenticeship is passing the NJATC aptitude test. A committee selects high-scoring applicants for interview.

Government agencies and higher learning facilities also offer apprenticeship programs. The government regulates the apprenticeship system. The U.S. Department of Labor oversees a Bureau of Apprenticeship and Training. Apprentices are employed under contract, based on the National Apprenticeship Act of 1937. Trade unions regulate apprenticeships in some states.

GETTING PAID TO LEARN

Today, electricians, like apprentices in other construction industries, earn pay during their training time. The definition of "apprentice" has changed over the centuries. During the Middle Ages, an apprentice was a child or teenager taken under the wing of a skilled craftsman or tradesman. The young person received room and board but no (or very little) pay. An apprentice performed the dull "dirty work" in the master's shop, while watching the master work and learning skills. The master, in turn, obtained very cheap assistance.

Apprentices today are more like rookie employees. Apprentice electricians work for licensed electricians as paid assistants. It's the "earn as you learn" principle. Unlike other careers that require a college degree, apprentices incur little to no student debt. They may be required to provide their personal hand tools and pay union dues, but this is substantially less than the cost of a four-year degree.

The Occupational Outlook Handbook observes, "Because of the comprehensive training received, those who complete apprenticeship programs qualify to do both maintenance and construction work."

Entering the profession as an apprentice is a career investment. In time, the apprentice becomes a professional—a journeyperson or journey worker. (The term journeyman is noninclusive and outdated.)

Students can find information about apprenticeships and other electrician job opportunities at their state employment offices or apprenticeship agencies. Local electricians can provide details. Students can also check out their state's department of labor website for apprenticeship information. Additional sources of guidance include the Independent Electrical Contractors (IECI), NECA, the Associated Builders and Contractors (ABC), the IBEW union and the Home Builders Institute (HBI). (See "For More Information" in the back of this book for contact information for these organizations.)

An apprentice works under the supervision of a licensed electrician. At the end of four or five years, they'll know how to do their job safely and correctly.

MILITARY TRAINING

Another way to become trained as an electrician without incurring student debt is through the armed forces. The Army, Navy, and Air Force all offer career paths involving electronics. Many of these options are open to full-time enlisted personnel as well as reservists and members of the National Guard.

Electricians in the military do many of the same things as civilian electricians—they make sure that the wiring in buildings is safe and functional, and that things like lighting systems are working. They might also do outside lineman work, ensuring that buildings are properly connected to the power grid. There are other exciting options, such as working in aircraft electronics or maintaining and repairing specific defense systems.

Specific Armed Services Vocational Aptitude Battery (ASVAB) scores are required for certain programs. Military training is generally counted as a pre-apprenticeship in the civilian world. The IBEW has programs to help veterans transition to civilian electrician careers.

ELECTRICIAN 101

One of the most compelling reasons to learn a trade is that you'll start your career with at least four years of hands-on experience and little to no student debt. An apprentice electrician will gain professional experience in the following aspects of the job:

- installation of equipment and wiring to specifications
- NEC and local building codes

- use of specific tools to repair equipment and electrical systems
- ability to troubleshoot problems—to find the causes and determine the best solutions
- ability to maintain equipment, keeping it in safe working order
- equipment and technology design and selection—deciding what equipment and tools are necessary for a job, and adapting them to provide satisfactory service
- operations analysis and monitoring—determining the requirements for an electrical project and, after installation, checking to ensure it is functioning properly
- financial management—getting the best value for money spent on an electrical project, and keeping accurate records on expenses
- proper use of PPE and appropriate first aid and safety practices

In addition, they'll have classroom study and homework in the following fields:

- **Building and construction:** The student learns about materials, methods, and tools that are required for the construction and repair of buildings, streets, and other structures. This includes the latest in "green" materials and construction methods.
- **Mechanics:** An electrician must be familiar with the use, design, and maintenance of tools and equipment that are used on the job.

- **Design:** An electrical student examines building blueprints, plans, and design sketches and models, and learns about the structural design process.
- **Production and processing of materials:** Electricians must be familiar with raw material production, quality-control issues, costs, etc.
- **Physics:** Physical principles affect most aspects of electricians' work—electrical, mechanical, material, and more.
- **Personal skills:** A successful electrician must be able to ensure customer satisfaction. This involves understanding the customer's needs, explaining the possible solutions in plain language, providing quality service, and evaluating customer satisfaction after the project is completed.

According to the description in the BLS Occupational Outlook Handbook, apprentices learn electrical theory, blueprint reading, mathematics, electrical code requirements, and safety and first-aid practices. They might also be trained in soldering, communications, fire alarm systems, cranes, and elevators.

Apprentices work under the supervision of experienced electricians. They begin with mundane tasks such as drilling holes, setting anchors, and attaching conduit. Later, they measure, fabricate, and install conduit and install, connect, and test wiring, outlets, and switches. They also learn to set up and draw diagrams for entire electrical systems. Eventually, they practice and master all of an electrician's main tasks.

Electricians need to be able to read blueprints and building plans. They have to understand how electrical systems work with plumbing, HVAC, and other aspects of building design and construction.

BEYOND THE BASICS

Beyond basic installation and repairs, there are a variety of careers for electricians. Some require special training in topics such as computer technology, industrial electronics, electromechanical and instrumentation technology, power transmission installation, line work, or quality control. Certain programs may require courses in topics such as mechanics, physics, home design, and customer service.

Here are some examples of training programs required for related professionals:

- **Electrical and electronics installer and repairer:** Often these professionals have studied electronics at a trade school or community college. They usually have a specialty, such as appliances. Electronics technicians may choose to pursue a professional certification from the Electronics Technicians Association International (ETA International) or the International Society of Certified Electronics Technicians (ISCET). Certification is dependent on meeting training and work experience prerequisites, as well as passing an exam.
- **Electrical and electronics engineers:** This field requires a bachelor's degree in engineering. Some universities have five-year programs culminating in both a bachelor's degree and a master's degree. Becoming a licensed or professional engineer requires a degree, a passing score on the Fundamentals of Engineering (FE) exam, relevant work experience, and a passing score on the Professional Engineering (PE) exam.
- **Telecommunications equipment installers and repairers:** Telecom technicians require special training in subjects such as electronics, telecommunications, and computer networking. An associate's degree or certificate from a trade school is a prerequisite for this field. Novices will also receive on-the-job training. Technicians who work with fiber optic cable may withs to pursue certification from the Fiber Optic Association (FOA).

- **Power line installer and repairer:** Installing and maintaining power lines requires extensive on the job training. Due to the hazards of the work, substance abuse testing is common, as is training in rescue work and emergency first aid. The ETA offers certification in this field.

THE NEC CODE

The basic codebook for electricians is the NEC. NECA calls it the "bedrock of the electrical construction business." It is revised every three years. Changes are based on new information and recommendations from professional electricians, consumers, product manufacturers, inspectors, and others.

The National Fire Protection Association (NFPA) publishes the NEC. The NFPA was established in 1896. Its goal is to eliminate death, injury, and property damage due to fire and electrical hazards. It develops and publishes the standards and codes for fire safety. They run public safety campaigns designed to educate the public on preventing and escaping fire.

The NEC is designed for professionals. It is more than 700 pages long, and electricians are advised to become very familiar with its contents. Electrical companies and unions frequently teach classes on NEC changes for their employees.

The NEC is not an instruction book. It's not intended to teach electricians how to do their work. Its purpose is to ensure that all their installations, upgrades, and repairs are performed according to the latest proven safety standards.

WHY DO WE HAVE ELECTRICAL CODES?

The Towering Inferno is an all-star blockbuster movie released in 1974. It's based on two novels, *The Tower* by Richard Martin Stern, and *The Glass Inferno* by Thomas N. Scortia and Frank M. Robinson. It's regarded as one of the all-time classic disaster movies in Hollywood history.

In the film, a fire breaks out in a high-rise building because the builder cut corners and installed cheaper electrical equipment than was adequate for the structure. Although this makes for a very dramatic story, it's not anything people want to encounter in real life. The movie was made with technical advice from the Los Angeles and San Francisco fire departments and outlined the need for proper wiring and adequate safety systems.

In a curious coincidence, a real fire occurred in São Paulo, Brazil, in 1974. It was caused by an electrical short in an air conditioning unit. The blaze killed 179 people, making it the worst high-rise disaster in history until the September 11, 2001, terrorist attacks.

Fire codes, building codes, and electrical safety codes such as the NEC exist to prevent tragedies like this.

The NEC is not "the law," either. The NFPA cannot enforce it. State and local governments, though, base their wiring regulations on the NEC. Municipal building codes are enforceable in court and violation can result in monetary fines.

Sad experience has proven that building safety is ultimately less expensive than a preventable calamity. If new electrical installations do not meet NEC standards, power companies usually will refuse to connect raw power to the building. Insurance companies will contest damage or injury claims that could have resulted from faulty wiring. They may refuse to insure the facility altogether unless the electrical work conforms with the NEC.

Other standards used by electricians include Building Officials and Code Administrators (BOCA) guides and the International Residential Code (IRC). Electricians must comply with federal OSHA standards and may need to pass an OSHA exam. Some state and local regulations impose additional requirements for electrical work.

NECA publishes its own set of codes, the National Electrical Installation Standards (NEIS). According to NECA, they "go beyond the minimum safety requirements of the National Electrical Code." Their intent is to ensure that electrical products and systems are installed in a "neat and workmanlike manner." The association explains, "With NEIS, the electrical installation you design or install not only meets code—it meets the shared expectations of everyone involved: owner, specifying engineer, electrical contractor, and the authority having jurisdiction."

SO MANY WAYS TO LEARN

Although learning a trade isn't a huge monetary investment, like studying medicine or law, a five-year apprenticeship is still a big commitment. Someone who's unsure that they want to be an electrician might take a job as an electrician's helper. Rather than performing electrical work, they can be useful in preparing work sites, running errands, helping gather and transport material, etc. As unskilled labor, they'll mostly be doing boring, dirty tasks, but they'll get to see electricians and apprentices in action and decide if it's what they really want to do.

Some electrical students might be very committed to the theoretical aspects of the job and complete much or most of their classroom study before they become on-the-job apprentices. Many vocational and technical institutions offer courses. Prospective apprentices with a degree, certificate from a trade school, or military training as an electrician begin their apprenticeships at a more advanced level than those without prior training.

The internet has made online learning opportunities available to millions of people who can't enroll in classes at a physical college campus. The COVID-19 pandemic has refined distance learning even more. Some institutions offer study-at-home training for aspiring electricians.

Penn Foster has been a pioneer in distance learning for more than 130 years. Originally, they were known as a correspondence school—that is, they mailed assignments and materials to students, who worked on their own and submitted tests for grading. Today, most of their programs are online. They have career school diploma programs for electricians, basic electronics trainees, and electronics technicians. The electricians program teaches such subjects as electricity classifications and categories; conduit fittings and supports; the installation of residential, commercial, and industrial electrical components; heating and lighting methods; and the properties of electrical conductors.

Note that students who successfully complete the programs are not yet authorized to enter the workplace as electrical and electronic service professionals. They must pass the required testing for certification, such as the Electronics Technicians Association Certification Exam, and acquire on-the-job training. Penn Foster's programs include exam preparation.

Penn Foster offers an associate's degree in electrical engineering technology. Studying at home, students take courses in such subjects as fundamentals of electricity, electrical and electronic measurement and instruments, and electric motors and controls. They ultimately work in the business sector, manufacturing, or government agencies.

The institution has an associate's degree program in electronics technology. Students learn electronics fundamentals, electronic instrumentation and measurement, the testing of circuits and electronic components, and industrial electronics. They may go to work for an electronics repair business.

As with any educational decision, you'll need to do your research. Are there any trade schools in your area? How about community colleges with suitable classes? How much do they cost? Your school's guidance counselor should be able to help you with these questions.

See if there are any local businesses you can visit. Is there a local electrician you can talk to? How about getting a summer job at an electrician's office or at an electrical supply store?

CHAPTER FIVE

YOUR FIRST JOB

After graduating high school, and possibly completing some additional course work or pre-apprenticeship training, an aspiring electrician needs to get real world experience. Most will acquire this training as apprentices. Guidance counselors can help you find apprenticeships in your area. You can also check out your state's department of labor website or your local union website. Apprenticeships are also listed on job search websites such as Indeed.

One of the most useful tools in a job search is a résumé. This is a document that tells potential employers about the applicant's educational background, work experience, and unique skills. When you're seeking your first job or apprenticeship your résumé will include your educational background, relevant extracurricular activities, and any special skills or training you might have, like speaking a second language fluently, or having a first aid or CPR certification. Often public libraries or school career offices have workshops on creating résumés.

Workers should update their résumés as they acquire new skills and advance in their jobs. Styles change over time and a good résumé is more than just a list of jobs and dates. When seeking a new job, it's always a good idea to refresh your résumé in the current format.

JOB CORPS AND THE HOME BUILDERS INSTITUTE

Job Corps is a U.S. Department of Labor program that prepares young people from low-income backgrounds for well-paid, meaningful careers. Job Corps is open to young people aged 16 to 24. Participants receive housing, meals, medical care, and a stipend. They also receive books and supplies and remedial education if needed. Job Corps can help students who need to finish high school. The program is federally funded, and participants don't incur student loan debt.

Job Corps counselors can help students who wish to attend a community college, enlist in the military, or apply for an apprenticeship. Job Corps offers training in several construction trades, including six types of electrical work. Some of their programs count as pre-apprenticeship training which can be applied to a formal union apprenticeship.

The Home Builders Institute (HBI) is another program funded by the Department of Labor. They provide free pre-apprenticeship training in construction trades. HBI works in partnership with the Job Corps. HBI's programs take about a year to complete. Students earn a pre-apprenticeship certificate and their OSHA-10 certification.

GETTING YOUR LICENSE

The next step is to become licensed. Without a license, you can't work as an electrician without supervision. Requirements for a license differ among states, but certain fundamental knowledge is the same everywhere. Applicants must prove they thoroughly understand electrical theory and know the NEC.

They also must know the requirements of the electrical and building codes imposed by their local government agencies. Once they pass their licensing exam, they're considered a journey worker.

The next level is master electrician. Master electricians are veteran electricians with at least seven years of experience; many have electrical engineering degrees. To become licensed as a master electrician, a candidate must have 4,000 hours experience as a journey worker, complete a training program, and pass an exam.

Although exact requirements vary by municipality, it's generally illegal to perform electrical work without a license. Even working on your own home or doing someone a favor is frowned upon. Electricity is simply too dangerous to trifle with. The system of classroom training, supervised OJT, and standardized tests exists to protect people and property from harm.

Someone trying to do electrical work without proper training might electrocute themselves or start a fire. Additionally, when a house is put up for sale, often prospective buyers hire inspectors. These inspectors know how to spot electrical work that's not up to code and can stop a sale until the work is corrected.

It's also common for even an experienced contractor to start a project, open up a wall, and realize there's a major issue that needs to be addressed. A licensed electrician will be able to assess the situation and use their training and experience to determine the best and safest way to proceed.

When renovating a house, it's not unusual for contractors to open a wall and discover serious electrical issues. Perhaps the wiring is simply old and outdated. Or it might have rodent damage. A licensed electrician can explain to the homeowner what must be done and why.

NATIONAL ELECTRICAL CONTRACTORS ASSOCIATION (NECA)

NECA was founded in 1901 by a group of electrical contractors at the Pan-American Exposition in Buffalo, New York. NECA's goal is to set standards for safety and training in the electrical industry. They offer educational programs designed to help contractors be better electricians and better businesspeople.

NECA is active in Washington, D.C., advocating for legislation that aids the electrical industry. Their Council on Industrial Relations (CIR) is available to help mediate labor disputes, in partnership with the IBEW. NECA offers a variety of classes and conferences that members can attend to learn about new industry advances and network with other members.

Their partner organization National Electrical Installation Standards (NEIS) describes the standards for correct installation of electrical systems. The NEIS tackled the job of defining in precise terms what counts as "good workmanship" for various electrical installation tasks.

These codes and standards exist to keeps both electrical contractors and the general public safe. Doing a job safely and correctly might appear to cost more, but it's ultimately cheaper and definitely safer than subpar work that needs to be redone, or worse—causes a fire.

BEYOND JOURNEY WORKER

It's a wonderful accomplishment to finish your training, pass your exam and finally obtain an electrician's license. Education and testing never end, though. Professionals must keep their licenses current. They may need additional courses from time-to-time concerning new types of electrical systems.

Meanwhile, professional electricians must stay abreast of changes in building codes and license requirements. First aid and CPR certifications need to be renewed periodically. Someone looking for professional advancement may need to upgrade their OSHA 10 certification to an OSHA 30—and OSHA recommends periodic refreshers courses for both, even though they're not generally required.

Manufacturers of electrical products often provide training for their electrician customers. Electrical contractors, or electricians who wish to become contractors, can benefit from business and management courses.

Advancement in the electrical professions is based on increasing experience. It's also affected by the changing workplace.

The BLS reports that almost 66 percent of electricians work for electrical contractors and wiring installation contractors. About 7 percent are self-employed. The rest work in manufacturing, for the government, or in employment services. Many young people who are considering a career as an electrician assume they will begin by working under experienced professionals, but eventually will start their own business.

The Occupational Outlook Handbook points to other advancement paths, such as the job of electrical supervisor. Experienced electricians who have years of experience in construction might diversify and become construction superintendents or project managers. This sort of job might require becoming licensed as a master electrician.

Major building projects can be both residential (multi-home neighborhoods) and commercial/industrial. Some project managers

Firefighters respond to thousands of home electrical fires each year. These fires cause hundreds of deaths and billions of dollars in property damage.

and superintendents who have shrewd business and management skills decide to launch their own contracting businesses.

General electricians can advance by becoming specialists. They may choose to study avionics, solar panel systems, or elevators and escalators. Or they might choose something less physical, such as drafting or designing electrical systems.

A different advancement opportunity is to become an electrical inspector for an industry, government agency, or consulting firm. All electricians can tell if a basic installation or electrical item meets safety standards; an inspector has further expertise to spot hazards and regulation violations that are not so obvious.

UNIONS

A labor union is an organization of people who work in the same profession. By acting as a group, unions can negotiate for fair wages, benefits, safe working conditions, and uniform work hours. Generally, union members warn more than non-union employees.

The union that represents electricians is the International Brotherhood of Electrical Workers (IBEW). The IBEW represents 820,000 workers and retirees in the United States and Canada. It's the largest American union in the AFL-CIO.

The earliest industrial use of electricity in the United States was for telegraph lines in the mid-19th century. The work was hard, dangerous, and didn't pay much. The first convention of the National Brotherhood of Electrical Workers (NBEW) was held in 1891 in St.

Louis. Its goal was to set fair working conditions and improve safety for people who worked with electricity. The union expanded to Canada and the name was changed to the IBEW in 1899.

The IBEW represents many types of electricians—in construction, manufacturing, utilities, and telecommunications. Electricians who work in the entertainment industry are represented by the International Alliance of Theatrical Stage Employees (IATSE).

BEING BILINGUAL

According to the BLS, about 30 percent of construction workers and 25 percent of electricians are Hispanic. There's a definite need for professionals who are fluent in both English and Spanish. Working with electricity demands precision and adherence to complicated instructions. Being able to communicate effectively with workers on a construction site helps keep everyone safe.

In a residential setting, customers tend to be more comfortable with a professional who speaks their language.

OWNING YOUR OWN BUSINESS

Many people think owning your own business is the American dream, but it takes careful consideration. If you're the boss, you're not just doing the electrical work that you've trained for, you're also responsible for billing, payroll, insurance, and a host of other details.

Take a good look around your community. Is there enough work for another electrician? What will your specialty be? Do you plan to work alone or hire other people?

Will you need to acquire an office or storefront? Or can you dedicate a room in your home for use as an office? What about another cellphone for business use? A post office box? Website and social media? All these things cost money and some take time to maintain.

Will you need to purchase equipment such as ladders? What about a truck or van? A clean, professional-looking vehicle is a good advertisement, but people get nervous if they see a disreputable van lingering in their neighborhood.

You also need excellent people skills if you want to run your own business. You need to communicate clearly with your customers—they need to trust that you're acting in their best interests, not just trying to run up the bill. If you hire other people, you need to treat them fairly, while also "being the boss." It's a lot to think about.

Since 1976, Habitat for Humanity has helped people around the world have safe, decent, affordable housing.

HABITAT FOR HUMANITY

Habitat for Humanity is a nonprofit organization that helps provide safe, affordable housing for people in need. Volunteers provide a variety of services, from building homes, to performing renovations for elderly homeowners, to disaster recovery. Habitat homes are designed to be energy efficient and utilize "green" or sustainable building methods.

Habitat volunteers are closely supervised by professionals to ensure everyone's safety. It's a great way to give back to the community, whatever your level of expertise.

Linda and Millard Fuller founded Habitat for Humanity. Habitat's guiding principle is the belief that every family has the right to a safe, decent place to live.

CHAPTER SIX

WHAT DOES THE FUTURE HOLD?

Training as an electrician is a wise career path. When people talk about sustainable energy and alternative fuels, they're talking about the sources of electricity. All the fully electrified buildings that exist with their appliances and other electronic equipment aren't going anywhere. There will be a need for electricians to install new systems and maintain old ones for the foreseeable future. The BLS estimates about 79,900 job openings for electricians per year for the next decade. Most of these job openings will be replacing electricians who retire.

Climate change is one of the biggest headlines of the 21st century. As new sustainable energy resources are developed, there will be a need for trained technicians to install and maintain them. This includes technologies such as solar panels and wind turbines.

Career prospects are bright for line workers. The BLS projects 23,500 job openings per year for the next decade. Due to the physical nature of the job, older line workers might decide to specialize in a different field of electrical work, and of course, people leave the workforce to retire.

Telecommunications technology is constantly being expanded and upgraded. Most public buildings have Wi-Fi—often two separate systems, one used for business operations and another for guests. Depending on the nature of the business, there might even be more than two systems for guests, such as a free Wi-Fi system for the general public, and separate, better grade of Wi-Fi that guests pay for, or is offered as a perk to certain guests. A large office building might have a system for building operations, free Wi-Fi for the building, and a separate system for each company that maintains offices there.

All these systems require wiring, routers, and other specialized equipment that's constantly being upgraded. For example, if one hotel brand offers a new superfast level of internet service in all its properties, other brands will swiftly follow. That's a lot of wiring!

Cell phone systems are constantly being upgraded as well. New levels of service are introduced to offer better connections, more security, and faster rates of data transfer. This requires new towers, and older ones, along with their equipment, are retired. All these systems require highly trained electricians to install, repair, and maintain them.

It seems like every time you turn around, cell phone companies are offering a new, upgraded model. That tiny phone that fits in your pocket requires an increasingly complex network to support all of its services and features.

EVERY BUSINESS HAS ITS UPS AND DOWNS

Even the most stable career path occasionally hits a few bumps in the road. The COVID -19 pandemic affected every industry, including construction. Many people employed by entertainment or recreational venues, including electricians, lost their jobs. New construction projects were delayed or canceled by the need to implement health and safety protocols, lack of supplies due to manufacturing delays, and staff illness. Large office complexes didn't require as much electrical maintenance, since people were working from home.

Some types of electricians had plenty of work as people needed to upgrade their home internet to accommodate working from home or distance learning for their kids. This is where good basic electrical skills come into play—with solid training, an electrician can adapt to changing business circumstances.

A CHAT WITH A VETERAN ELECTRICIAN

Jerry Van Liew was a hospital electrician in Charleston. This is what he had to say about his career:

"I've been doing electrical work for 28 years and find it to be a very rewarding career. The satisfaction for me is knowing everything in our facility is working, and our patients and staff are happy."

The work requires constant alertness, though. “The thing I don’t like is working on live electrical circuits, which can be hazardous if you are not a trained, qualified electrician,” he says.

Still, he recommends an electrician career for any young person who has a mind for solving problems. By making sure electrical systems are reliable and safe, electricians perform a tremendous service to society.

Van Liew agrees it is a “recession-proof” career path. “There will always be a need for electricians,” he says.

Nationwide, all workers on construction projects, including electricians, are affected by economic and social changes like the pandemic. Workers had to figure out how to adapt their work PPE—like safety glasses—to work with medical PPE like masks, which make glasses fog up. They had to follow government health and safety protocols, as well as their client’s. Projects often had to be rearranged to accommodate the supplies and staff available.

Even when it’s not possible to build new structures, existing ones require maintenance. During the pandemic, commercial electrical use dropped substantially, but residential use rose. This meant that different systems were suddenly receiving much more use than usual.

Today, most pandemic restrictions have been lifted and there are new construction projects. Travel and entertainment venues are operating at full capacity. However, there are still difficulties obtaining some types of electrical equipment and supplies.

DOMOTICS

Not too long ago, if you wanted to change the setting on an appliance, you either had to get up and change it by hand or perhaps use a remote control that worked only at fairly close range. The idea that you could control many household systems from your computer, phone, or by voice was the stuff of science fiction—or the homes of the very wealthy.

Today domotics, or home automation, is becoming much more common and affordable. Smart home "hubs" such as the Google Nest or Amazon Echo use your home's Wi-Fi and your personal mobile device to control lighting, security and locks, and home entertainment.

Stoves, refrigerators, washers and dryers, HVAC equipment, and a host of small kitchen appliances come with built in Wi-Fi equipment. You can set up security cameras around your home and check on them via your phone when you're away.

All these complex electronic devices require trained technicians to service them, and reliable high-speed internet systems to operate.

A small hub like this can control many different appliances throughout a home.

LIVING THE DIGITAL LIFE

The use of personal computers and the internet has increased at an explosive rate during the past 25 years. Home and office consumers want the latest and greatest digital products. These include state-of-the-art entertainment systems, gaming systems, surveillance and alarm systems, and backup power systems that will keep their equipment running whenever normal power is interrupted.

People carry smart phones, laptop computers, e-readers, and small tablets with them everywhere. They expect to be able to connect to the internet in any public place and have easy access to electrical outlets to charge their devices.

Since the COVID-19 pandemic, many people still work from home full or part time and need reliable high-speed internet connections for videoconferencing. They've also come to rely on these systems for social functions and remote learning.

Controlling all those smart appliances from your watch? Yup, there's an app for that. All these gadgets require reliable internet and electricity, ensuring plenty of work for electricians.

Your love of electronics and gadgets can be parlayed into a well-paying career as an electrician.

The digital lifestyle influences home design as never before. The modern home is a place for high-fidelity entertainment, advanced communications, and automated convenience that rivals a corporate office building. That means modern wiring projects may involve whole-home remote controls, wireless internet systems, computer networks, public address systems, multi-room sound, and back-up power. Such wiring may extend outside to decks, the poolside, and outbuildings.

Electricians who learn how to wire homes and offices to meet the new demands of the digital generation have an advantage in marketing their services.

SPECIALIST OR GENERALIST?

Becoming an electrician is a solid career path with many specialties to choose from. Green building techniques, sustainable energy sources, and home automation are all exciting new fields with lots of job potential. However, the recent pandemic demonstrated that having solid basic skills and a flexible personality are very useful traits. Even when so many jobs were disrupted, there was plenty of systems that needed maintenance and repair. Hospitals, with all their complex equipment, needed to be functioning at 100 percent. So did internet service providers and electrical utilities.

An electrician with solid professional skills, a good understanding of the NEC, and an open mind can always find work.

Whether you want to work on new construction projects or be the guy your neighbors turn to when their lights go out, becoming an electrician is a solid career choice.

10 GREAT QUESTIONS TO ASK A LICENSED ELECTRICIAN

Here are some questions you can ask a licensed electrician. If there's a specialty you're interested in, see if you can talk to someone in that field. Otherwise, talking to a local residential electrician is a great place to start.

1. Did you go to college or a trade school?
2. Can you recommend a school nearby? About how much will it cost?
3. Are there any high school classes or activities you recommend?
4. Do you have a specialty? Why did you choose it?
5. Do you like working on your own or as part of a team?
6. Are there any books you recommend I read? Industry blogs or magazines that are worth following?
7. Has your job changed because of concerns about sustainable energy sources?
8. Did you go back to school to learn about new technology?
9. Do you have any certifications? Which ones would you recommend I get?
10. Is there anything you wish someone had told you when you were considering this career?

What other questions can you think of?

A science museum is a fun place to learn about electricity. Who knows—maybe one day you'll be the one designing the exhibits!

GLOSSARY

academic: Pertaining to school studies.

analytical: Requiring a problem to be separated into parts to find a solution.

blueprint: A print of a technical drawing with white lines printed on a blue background, or a print with blue lines on a white background, usually of an architectural or engineering design.

cardiopulmonary resuscitation (CPR): The emergency treatment of a person whose heart has stopped beating.

circuit: A wiring system that supplies a group of outlets with power.

circuit breaker: A device that automatically opens (disables) an electrical circuit when it becomes overloaded.

codes: A set of rules and requirements.

component: A part.

conduit: A protective, nonmetallic sleeve or "raceway" through which electrical wires are run.

consumer: A buyer and user of a product or service.

dexterity: The easy use of hands, arms, and legs.

digital: The functioning of a device (such as a computer) by numerical units.

electrocution: Death by electrical shock.

flammable: Easy to ignite.

grounding: Returning electrical current to the earth.

insulation: Material that prevents an electrical wire or cable from causing a shock or fire.

preventative maintenance: Electrical replacement or upgrading done for the purpose of avoiding future disasters.

radar: An electronic system that uses radio waves to detect and identify distant objects.

receptacle: A mounted electrical fitting that contains the live parts of a circuit.

regulations: Rules established by a government agency or other legal authority.

résumé: A short summary of a person's educational and work experience.

schematic: The diagram of an electrical system.

SONAR: An underwater electronic detection system.

standards: A body of procedures and rules that have become accepted over time.

telecommunications: Communications over long distances by telephones, now expanded by the internet and satellite systems.

transformer: An electrical component that alters voltage levels.

utility: Electrical, plumbing, heating, or a similar service that provides daily needs.

videoconferencing: People in distant locations meeting online, in real time, via computers and the internet.

FOR MORE INFORMATION

ASSOCIATED BUILDERS AND CONTRACTORS

440 1st St., NW, Suite 200
Washington, D.C. 20001
(202) 595-1505
Website: abc.org
Facebook: @ABCNational
Instagram: @abc_national
ABC is a national construction industry trade association founded in 1950.

BUREAU OF LABOR STATISTICS

Postal Square Building
2 Massachusetts Ave., NE
Washington, D.C. 20212-0001
(202) 691-5200
Website: bls.gov
The BLS is an official website of the U.S. Department of Labor. Their website all sorts of details and statistics for a multitude of career choices.

ELECTRICAL SAFETY AUTHORITY

155 Matheson Blvd. West
Mississauga, ON L5R 3L5
Canada
(905) 712-7819
Website: esasafe.com
Facebook: @ElectricalSafetyAuthority
The ESA is responsible for regulating and promoting electrical safety in Ontario.

ELECTRICAL TRAINING ALLIANCE

5001 Howerton Way, Suite N
Bowie, MD 20715
888-652-4007
Website: electricaltrainingalliance.org
Facebook and Instagram: @ibewnecaeta
ETA is the leading voice in education for the electrical industry.

EMPLOYMENT AND TRAINING ADMINISTRATION

U.S. Department of Labor
Frances Perkins Building
200 Constitution Ave. NW
Washington, D.C. 20210
(877) US-2JOBS [872-5627]
Website: www.doleta.gov
The ETA oversees job training and employment-assistance programs, which are generally administered at a local government level. It's part of the U.S. Department of Labor.

HOME BUILDERS INSTITUTE

1201 15th St. NW, Sixth Floor
Washington, D.C. 20005
(202) 371-0600
Website: hbi.org
Facebook and Instagram: @HBIBuildingcareers
The HBI provides free pre-apprenticeship training in a variety of trades.

INDEPENDENT ELECTRICAL CONTRACTORS

2900 S. Quincy St., Suite 720
Arlington, VA 22206
(703) 549-7351
Website: ieci.org
Facebook: @IEC.national
Instagram: @IEC_National
IEC is a trade organization for the electrical industry.

INTERNATIONAL BROTHERHOOD OF ELECTRICAL WORKERS

900 Seventh St. NW
Washington, D.C. 20001
(202) 833-7000
Website: ibew.org
Facebook: @IBEWFB
The IBEW is the trade union that represents electricians in the United States and Canada.

NATIONAL ASSOCIATION OF HOME BUILDERS

1201 Fifteenth St. NW
Washington, D.C. 20005
(800) 368-5242
Website: nahb.org
Facebook: @NAHBHome
Instagram: @homebuilders
The HBA works to improve workforce development in the construction trades.

NATIONAL ELECTRICAL CONTRACTORS ASSOCIATION

1201 Pennsylvania Ave. NW, Suite 1200
Washington, D.C. 20004
(202) 991-6300
Website: necanet.org
Facebook: @necanet
NECA advocates for safe and innovative construction in the United States.

NATIONAL FIRE PROTECTION ASSOCIATION

1 Batterymarch Park
Quincy, MA 02169-7471
(617) 770-3000
Website: nfpa.org
Facebook: @theNFPA
Instagram: @nfpadotorg
The NFPA is a nonprofit agency dedicated to fire safety and prevention. They publish the NEC.

FOR FURTHER READING

Carosso, Juan. *At Your Best as an Electrician: Your Playbook for Building a Successful Career and Launching a Thriving Small Business as an Electrician (At Your Best Playbooks)*. New York, NY: Skyhorse Publishing, 2018.

Clark, Brandon. *Journeyman Electrician Exam Prep 2023-2024 Version: The Ultimate Study Guide to Pass the Exam on Your First Try | Test Simulations, Answer Keys & Tips to Score a 99% Pass Rate*. Brandon J. Clark, 2023.

Hudson, Dunstan. *Wiring For Beginners 2023: The Complete Guide to Demystify Home Wiring and Residential Electrical Systems with Easy-To-Understand Language & Step-By-Step Pictures*. Independently published, 2023.

Lowe, Doug. *Electronics All-in-One for Dummies 3rd Edition*. Hoboken, NJ: John Wiley & Sons, Inc., 2022

Mapua, Jeff. *A Career as an Electrician (Jobs for Rebuilding America)*. New York, NY: Rosen YA, 2019.

Miller, Charles. *Ugly's Electrical References, 2023 Edition 7th Edition.* Burlington, MA: Jones & Bartlett Learning, 2023.

National Fire Protection Association (NFPA). *National Electrical Code, 2023 Edition, Spiral-bound with Tabs.* Quincey, MA: National Fire Protection Association (NFPA), 2022.

Platt, Charles. *Make: Electronics: Learning by Discovery: A hands-on primer for the new electronics enthusiast.* Sebastopol, CA: Make Community, 2021.

Santore, Marcia. *Electricians (Practical Career Guides).* Lanham, MD: Rowman & Littlefield, 2021.

INDEX

ABOUT THE AUTHOR

KATHLEEN A. KLATTE is the author of many nonfiction books for children and teens. Topics range from animals and nature to constitutional law to unusual career choices. She lives in New York with one cat and far too many books and Legos.

PHOTO CREDITS

Photo credits:Cover, p.1 sturti/iStock.com; p.3-96 Merydolla//Shutterstock.com; p. 5 Amorn Suriyan/Shutterstock.com; p. 6 Andrewshots/Shutterstock.com; p. 7 REDPIXEL.PL/Shutterstock.com; p. 8 Zenobillis/Shutterstock.com; p. 9 Ground Picture/Shutterstock.com; p. 10 claudio zaccherini/Shutterstock.com; p. 11 GOLFX/Shutterstock.com; p. 13 Uwe Aranas/Shutterstock.com; p. 14 Golubovy/Shutterstock.com; p. 15 Pixel-Shot/Shutterstock.com; p. 18 ungvar/Shutterstock.com; p. 20 M Stocker/Shutterstock.com; p. 25 Quality Stock Arts/Shutterstock.com; p. 27 kittirat roekburi/Shutterstock.com; p. 30 Tim Daugherty/Shutterstock.com; p. 31 RossHelen/Shutterstock.com; p. 33 FOTOGRIN/Shutterstock.com; p. 35 John Penezic/Shutterstock.com; p. 37 TungCheung/Shutterstock.com; p. 39 Svjatoslav Andreichyn/Shutterstock.com; p. 40 Pixel-Shot/Shutterstock.com; p. 43 Reimar/Shutterstock.com; p. 43 Aleksandr Grechanyuk/Shutterstock.com; p. 45 PeopleImages.com - Yuri A/Shutterstock.com; p. 47 Bannafarsai_Stock/Shutterstock.com; p. 49 Fit Ztudio/Shutterstock.com; p. 52 Chinnapong/Shutterstock.com; p. 53 skipper_sr/Shutterstock.com; p. 57 AUUSanAKUL/Shutterstock.com; p. 67 szefei/Shutterstock.com; p. 69 Dale A Stork/Shutterstock.com; p. 72 Kathy Hutchins/Shutterstock.com; p. 73 Casimiro PT/Shutterstock.com; p. 75 Ivelin Denev/Shutterstock.com; p. 78 Gorodenkoff/Shutterstock.com; p. 79 DenPhotos/Shutterstock.com; p. 80 Ground Picture/Shutterstock.com; p. 81 Lopolo/Shutterstock.com; p. 83 Zyabich/Shutterstock.com.